Once There Was a Deep Forest

By Jim Tarvin

Illustrations by Tessa Schaumburg

Edited by Lydia DeFusto

ISBN: 979-8-9885820-0-7

For addition copies of this book, visit Amazon.com

Dedicated to Liz

And to Our Children
Jimmy, Michael, Peggy, and Elizabeth

Preface

I have tried to preserve the amazing beauty of the forest with the words and paintings on these pages.

I hope this book will inspire you to go into the forest and appreciate its plants and animals. Maybe you, too, will write about or create images of what you see.

Will these forests always exist? They could if we protect them by keeping our air, land, and water clean.

It's up to all of us.

~ Jim Tarvin

A beautiful stream flowed through the forest.

Frogs and sunfish and beaver dotted the pond.

As Fall arrived, the sun got lower and lower in the sky.

On crisp nights, the stars blinked brightly.

Turtles sunned themselves on half-sunken logs.

Robins sat on blue eggs in their nests.

They soon had many mouths to feed.

The circle of the seasons and the circle of life had begun again in the deep forest.

NOT the End

It's summer again!

Go back to the first page
and start over - just as the seasons do.

About the Author

Jim Tarvin, a retired college English professor from Goshen, N.Y., attributes his love of nature to his dad, Everett Roy Tarvin. Jim recalls the monthly arrival of *The Conservationist,* an environmental magazine. Everett was an avid birder and hiker who helped clear the original path of the Appalachian Trail. His grandson and namesake, Michael Everett Tarvin, is now section hiking the Trail. Jim hopes people who read this book love and want to protect our fragile world with its beautiful forests, streams, ponds, mammals, insects, fish, and reptiles. For children, the story is an invitation to the natural world.

About the Illustrator

Tessa Schaumburg grew up in Goshen, N.Y. and currently lives in Kerhonkson surrounded by the gorgeous Shawangunk Mountains. A graduate of SUNY New Paltz, Tessa studied visual arts. Along with artistic pursuits including illustrating, painting and pottery, Tessa works for a collie breeder caring for puppies and the many dogs on the farm. Nature is an integral part of Tessa's life and her main source of creative inspiration. She passionately loves all the critters of the forest. She urges everyone to spend time appreciating and protecting our planet and the beautiful animals who inhabit it.

Made in the USA
Middletown, DE
09 October 2023